Dilli

by

Donna Butterworth

Dedicated to the Donna Butterworth Gebler
grand and great-grand children

"Dilli"

Dilli was a daffodil bulb. She lived deep, down in the earth where it was warm and cozy and away from Mr. Northwind and Mrs. Snowflake.

But, Dilli was very unhappy.
She didn't like being a
daffodil bulb. She wanted
to be a flower. And, every
day, as Mr. Northwind blew
and Mrs. Snowflake fell,

Dilli said:

"Fee, fi, fiddle dee dee
A dancing daffy-dilly
will be me."

One day as Dilli was saying

"Fee, fi, fiddle dee dee

A dancing daffy-dilly

will be me",

Mr. Northwind changed to

Mr. Southwind, and

Mrs. Snowflake melted.

Dilli's warm, cozy house deep in the earth got warmer and warmer. So, Dilli decided to see what was happening on earth.

Dilli pushed her green head up further, further, and further until.... It poked right out of the ground. What do you think Dilli saw?

She saw a blue sky, budding trees, and singing robins.

It was spring !!

Dilli smiled and said,

"Fee, fi, fiddle dee dee
A dancing daffy-dilly
will be me."

Each day, as Mr. South-
wind gently blew and
the sunshine dazzled, Dilli's
head pushed higher and
higher into the sky until she
became a stem with a
leaf on each side.

One day the wind was stronger, and the sun was hidden by black clouds. Soon it began to rain. Dilli liked the rain because it helped her to grow.

But, she didn't like the worms that wriggled around her stem after the rain. Later she learned that the worms were help-ing her to become a flower.

Do you know how
they helped ?

By wriggling around her
stem, they loosened the
soil so that she could
easily drink the water.

Soon a bud appeared on Dilli's stem. She was so excited that she said,

"Fee, fi, fiddle dee dee
A dancing daffy-dilly will be me."

And, the bud grew and

grew until one morning

Dilli blossomed.

She was a beautiful

bright yellow

daffodil.

Dilli nodded to the robins, to the worms, and to the sun; and she danced in the breeze from Mr. Southwind.

Now, she was the happiest flower in the garden.

The
End

Follow Dilli the daffodil bulb on her journey as she becomes
what she was meant to be, a beautiful, happy flower blooming in
the garden. Donna Butterworth wrote and illustrated 'Dilli' in
1953. At this time, she was attaining her BS degree in early
childhood development from the University of Utah. The
illustrations and the story text have been printed to replicate their
original hand-drawn form.   Although it has been more than 70
years since Donna created the story of 'Dilli' the vintage style has
a simple yet delightful and timeless appeal. Donna loved all
flowers but had a special love for the daffodil. She would let all
know that the bright cheerful yellow flower represented
everything hopeful about springtime. It was not until after Donna
Butterworth's passing that the original manuscript of 'Dilli' was
discovered in a box of her memorabilia. Donna Butterworth
would be thrilled to know her storybook has finally been
published. She would want all children young and old to find joy
in their own personal journey and in discovering their potential.